— OUTDOOR ADVENTURES —

# FISHING

By Tom Carpenter

SportsZone

An Imprint of Abdo Publishing
abdobooks.com

**abdobooks.com**

Published by Abdo Publishing, a division of ABDO, PO Box 398166, Minneapolis, Minnesota 55439. Copyright © 2020 by Abdo Consulting Group, Inc. International copyrights reserved in all countries. No part of this book may be reproduced in any form without written permission from the publisher. SportsZone™ is a trademark and logo of Abdo Publishing.

Printed in the United States of America, North Mankato, Minnesota
092019
012020

Cover Photo: Andy Bowlin/iStockphoto
Interior Photos: Dan Thornberg/Shutterstock Images, 5, 23 (walleye); iStockphoto, 7, 13, 15, 23 (hooks), 25, 26, 29, 35, 36, 41, 42; Mindscape Studio/Shutterstock Images, 9; Dudarev Mikhail/Shutterstock Images, 10; Christopher Babcock/ Shutterstock Images, 16; Shutterstock Images, 19, 20, 23 (trout), 23 (pike); Steven Russell Smith Ohio/Shutterstock Images, 23 (panfish); Vincent Noel/Shutterstock Images, 23 (bass); Oleksandr Lytvynenko/Shutterstock Images, 23 (catfish); JLF Capture/iStockphoto, 30; Dieter Meyrl/iStockphoto, 33; Michelle Lee Photography/Shutterstock Images, 39; Cindy Creighton/Shutterstock Images, 44–45

Editor: Patrick Donnelly
Series Designer: Colleen McLaren

**Library of Congress Control Number: 2019942005**

**Publisher's Cataloging-in-Publication Data**

Names: Carpenter, Tom, author
Title: Fishing / by Tom Carpenter
Description: Minneapolis, Minnesota : Abdo Publishing, 2020 | Series: Outdoor adventures | Includes online resources and index
Identifiers: ISBN 9781532190483 (lib. bdg.) | ISBN 9781532176333 (ebook)
Subjects: LCSH: Fishing--Juvenile literature. | Angling--Juvenile literature. | Recreational fishing--Juvenile literature. | Outdoor recreation--Juvenile literature.
Classification: DDC 799.1--dc23

# TABLE OF
# CONTENTS


**CHAPTER 1**
## WHY WE FISH .................................................. 4

**CHAPTER 2**
## WHERE TO FISH .............................................. 8

**CHAPTER 3**
## EQUIPMENT ................................................... 14

**CHAPTER 4**
## TYPES OF FISH ............................................... 24

**CHAPTER 5**
## BAIT .............................................................. 34

**CHAPTER 6**
## SAFETY TIPS ................................................. 40

GLOSSARY ......................................................... 46
MORE INFORMATION ........................................ 47
ONLINE RESOURCES ........................................ 47
INDEX ................................................................ 48
ABOUT THE AUTHOR ....................................... 48

# WHY WE FISH

It's early on a summer morning. Sunshine peeks through the trees. Two anglers step into a canoe and put on life jackets. One sits up front. The other pushes off and sits in the back. They paddle the canoe with the small river's current.

Birds sing in the trees. Frogs croak on shore. A butterfly flutters across the water. The water swishes and swirls. The man in front casts his fishing line, baited with a nightcrawler, into the water. Soon he feels a tug on the line—a fish is biting.

The angler reels up the slack in his fishing line, feels the weight of the fish on the other end, and sets the hook.

Fishing can be a great way for families to spend time together.

*Whoosh!* A smallmouth bass, hooked in the jaw, leaps out of the water. Then the fish dives deep. It fights hard. But the angler wins, hauling the bass into the boat. He holds the fish while his partner takes a photo. Then they release the bass back into the water.

Long ago, people used to fish just to eat. Some still fish for a living. They mostly use nets. But many people also fish for fun and adventure. And fishing is a great way to get outside and experience nature.



## MANY REASONS

People who love fishing enjoy the whole process. Anticipation builds as they pack their gear and travel to their favorite fishing spot. They look forward to getting a bite and fighting the fish as it pulls and jumps on the end of the line.

Spending time alone with a line in the water can be very relaxing.

Fishing is a great way to spend time with family and friends. Many traditions center around fishing. Being out on the water can help people feel happy and relaxed. It's easy to talk and tell stories while fishing. Fishing helps people escape their everyday world and get outdoors into nature. They can engage their senses with all that surrounds them. People can see birds in the grass, bugs in the air, and deer on shore. The swish of water and the rustle of a gentle breeze are pleasing to the ear. And the sweet smell of flowers and grasses on land and earthy scent of lake water fills the air.

# WHERE TO FISH

Freshwater fishing happens on many bodies of water. Some people fish in large bodies of water, such as the Great Lakes, which are even bigger than some states. Other spots, such as a tiny brook gurgling through a meadow, are so small that a person could jump across it. Good anglers know that understanding fish habitats is the first step to fishing success. Every type of water is different and calls for a specialized approach.

## LAKES

A lake is a body of water where there is little to no current. Some lakes are natural, formed by glaciers or other processes in nature. Many natural lakes are fed by springs. Some are fed by rivers or creeks.

*Different bodies of water favor different types of fishing.*

Some lakes are good for catching bass and bluegills, others for walleye and perch, and others for trout.

Some lakes are created by human activity. Most often that activity is damming a stream or waterway so the water backs up and forms a lake or reservoir. These kinds of lakes can have the same types of fish that natural lakes do, as long as they offer the appropriate habitat.

Different parts of a lake offer better opportunity to land a fish. Weed edges, beds of weeds, and drop-offs—places where the water drops from shallow to deep very quickly—are usually populated by fish. Anglers also can have luck fishing near rocky points, shorelines, and reefs or bars that rise out of the watery depths. These places attract baitfish and other forage and offer fish habitat variety.

## PONDS

Ponds are similar to lakes, but smaller. Anglers love ponds for two reasons. First, ponds are easy to fish because the surface area of the water is so small. There are only so many places the fish might be located. Also, ponds often have a very fertile habitat, providing enough food for large numbers of fish.

As with lakes, certain spots in ponds are more likely to provide opportunities to catch fish. They include the edges of weedy areas, the deep spot near a dam

## CREEKS AND STREAMS

Creeks and streams are smaller versions of rivers. Like pond fishing, creek fishing is fun because it is simple. Some creeks are good for cold-water species such as trout. Other creeks support warm-water species such as panfish and bass. Good places to fish in creeks include riffles of fast and shallow water, the pools below them, deep bends, and undercut banks and tree roots where fish hide from overhead predators.

## RIVERS

Some anglers like to fish in rivers. Rivers consist of moving water. That can make fishing challenging. But it can also help anglers find the spots where fish are hanging out. Good places to fish in rivers include bends, eddies where the water swirls, obstructions that slow or split the current, and spots off to the side of the main flow.

Some rivers are very cold and support fish such as trout. Other rivers have cool to warm water and are good for species such as smallmouth bass, walleye,

Rivers can be excellent places for fly-fishing.

and white bass. Catfish are especially popular among river anglers.

Big rivers have backwaters or overflow areas where water flows when the river is over its banks. Other backwaters are connected to the main river by side channels. Backwaters can be extremely good fishing places because they are shallow and fertile with lots of cover and forage.

# EQUIPMENT

Successful anglers know that good equipment is essential for fishing success. But just one fishing setup won't do. Different kinds of fish and fishing call for different kinds of gear—rods and reels, fishing line, terminal tackle, bait, and other accessories.

Rods and reels are essential fishing equipment. The reel holds the fishing line and is attached to a rod. Rods help anglers cast their bait and then fight a fish after it is hooked. There are three main types of reels—spincasting, spinning, and baitcasting.

Spincasting reels sit on top of the rod and are operated with a push button for casting out line. Spincasting reels are very reliable. Many beginners

Anglers have countless options when choosing their equipment.

Anglers fishing in big waters will want longer, sturdier rods.

like them because they are simple and easy
to operate.

Spinning reels sit under the rod and are a little
more challenging to use. Spinning reels cast line
farther and with more precision because anglers

use their fingers to control where the line drops into the water.

Baitcasting reels sit on top of the rod and are very challenging to use. A baitcasting reel has an open or free spool that helps increase the distance of the cast. Instead of pushing a button, anglers use their thumb to control the line. Baitcasting reels can be extremely accurate for casting.

Rods come in many varieties as well. Those used for salmon or muskellunge—also known as muskies—on big waters are long and heavy. Anglers fishing for panfish in ponds or for trout in creeks will want a rod that's short and light. Rods in the middle are good for catching bass and walleye. All rods have guides from reel to tip that hold the line. That keeps it from breaking away from the rod.

Graphite rods are most popular with professional anglers. Graphite is lightweight but strong and flexible. Fiberglass rods cost less than graphite. At one time, bamboo was a common rod material.

# LINES AND KNOTS

Fishing line is critical. When an angler hooks a fish and fights it, fishing line is the only connection from fish to human. Monofilament line is strong and hard for fish to see. Good quality monofilament line is affordable too. Those qualities make it the choice of many anglers.

Braided lines are also popular. These lines are extra strong and have little stretch in them, making it easier to feel even light bites. But braided lines are slippery compared to monofilament, so strong knots are essential.

Experienced anglers have learned—often the hard way—that tying strong knots is a key fishing skill. A poor knot between line and hook or lure, or line and leader, is a weak link that can come apart and let a fighting fish get away.

Good knots for tying line to the lure or hook include the improved clinch knot, Palomar knot, and

loop knot. Good knots for tying two lines together include the uni knot and blood knot.

## TERMINAL TACKLE

Terminal tackle is attached to the fishing line. It includes hooks, sinkers, floats, swivels, and leaders.

A loaded tackle box gives anglers many options.

Hooks hold the bait. A fish is hooked when it bites and the hook is set. Fishing hooks can be used with live bait, but artificial baits known as soft plastics are often attached directly to fishing hooks too. Artificial lures have their own hooks attached.

Sinkers go onto the fishing line to help carry the bait down to the fish. Some sinkers are made of lead. But ducks, geese, loons, and other waterfowl can get lead poisoning when they swallow lost lead sinkers. That's why today's anglers often use sinkers of other nontoxic material, such as tin, bismuth, and tungsten.

A float, also called a bobber, keeps bait from settling on the bottom. The float jiggles and often drops below the surface of the water when a fish bites. Most floats are made of foam or plastic. Fixed floats attach to the fishing line with a clip. Slip floats adjust for different fishing depths.

Sometimes anglers don't use floats. For this type of fishing, the angler often sets up a rig using a hook, sinker, and swivel. A swivel prevents the line from twisting and snarling.

Some fish, such as pike and muskellunge, have sharp teeth that can easily slice any fishing line. So pike and muskie anglers use steel leaders instead of tying the hook or lure directly to the fishing line.

## BOATS, MOTORS, AND MODERN ELECTRONICS

Some anglers use boats and motors to get out onto the water of lakes and rivers. For smaller waters, a boat might be 14 to 16 feet (4 to 5 m) long, with 10- to 25-horsepower motors. On big waters such as the Great Lakes, boats might be 30 feet (9 m) long with motors offering hundreds of horsepower.

Many of today's boats are equipped with modern electronics—sometimes called fish finders—that measure the depth of the water. They also mark structures such as rocks, weeds, bars, reefs, drop-offs, and ledges—all places where fish hang out. Some electronics even point out individual fish in the water.

## OTHER ACCESSORIES

Anglers need a variety of accessories for a successful outing. They store and organize their gear in tackle boxes and tackle bags. A landing net helps scoop up fish that have been reeled in. Some anglers wear vests that store tackle and gear. Many use containers such as aerated buckets, ice-lined coolers, and insulated boxes to carry bait and keep it alive. Good boots or waders are important for shore anglers or for anglers who stand in the water to fish.

# — RIGHT HOOK —

Hooks come in different sizes. The bigger the number, the smaller the hook. Here are common fish hook sizes and which fish they are often used to catch.

# CHAPTER 4

# TYPES OF FISH

Part of what makes fishing fun is the variety of fish swimming in lakes, ponds, rivers, and streams. Some gamefish are big, sharp-toothed predators that dart out from a hiding spot and attack their prey. Other gamefish are spunky little hunters that hide near cover to avoid being eaten themselves.

Some anglers like to go fishing for anything that bites and just see what ends up on the end of the line. Other anglers specialize in fishing for a particular species of gamefish. Those anglers might fish for the same fish every time out all season long, or they might target different fish on a different day.

The best anglers work hard to understand the kind of fish they're trying to catch. Knowing about a fish's

Anglers can target specific fish or just drop a line in the water and see what bites.

habitat and habits—where that fish lives, how it hunts, and what it eats—are the keys to fishing success.

## TROUT

Trout are sleek, streamlined fish that can live in only clear, cold water. Brooks, creeks, streams, and rivers with cold water make the best trout habitat. But trout can also live in ponds and lakes where the water is cold enough.

In a small brook, a four-pound (1.8-kg) trout might be a trophy. In a huge lake, it might take a 10- or even 20-pound (4.5- to 9.0-kg) trout to be considered a trophy. There are many species in the trout family, but three are the most common.

Brook trout are colorful fish that live in small streams and cold-water ponds and lakes. Brown trout live in streams of all sizes, but they also inhabit big lakes, even the Great Lakes. Browns are very wary and are challenging to catch.

Rainbow trout live in streams of all sizes. Rainbows like fast water. Rainbow trout are shiny and silvery, with a pink or red stripe along their sides. They are known for leaping out of the water when hooked.

## PANFISH

Panfish are called that because their shape is round like a pan. They don't get very big. In most places, a one-pound (0.5-kg) panfish is a trophy. But they are easy and fun to catch. Panfish bite willingly, tug hard, and are great to eat.

## WALLEYE

Walleye are predators that eat other fish, crayfish, leeches, and insects hatching in the water. Walleye will bite at any time of day, but many anglers concentrate their fishing effort at low-light times (early morning or late evening) and at night. Walleye can be found in rivers and lakes alike.

Saugers are closely related to walleye. Saugers look a lot like walleye, but they are more blotched or spotted. Saugers are river-oriented fish. They tend to stay deeper than walleye.

Bluegills are one of the most common panfish. Bluegills are common in the western United States and in other parts of the world. They are a type of sunfish. Sunfish usually live in or near weed beds, lily pads, docks, or rocks. They are colorful and need to hide from bigger predators. A typical sunfish might weigh a half pound (0.25 kg).

Crappies are another popular panfish. Crappies are most prevalent in the eastern United States. Crappies prefer weed beds for habitat, but these fish also venture out into big, open water to hunt minnows.

## BASS

Bass fishing is extremely popular across the country. Bass grow to good sizes. Common types of bass include the largemouth and the smallmouth. Depending on where it is caught, a five- to 10-pound (2.3- to 4.5-kg) largemouth is considered a trophy. With smallmouth bass, five pounds (2.3 kg) is a trophy in most places. A one- to three-pound (0.5- to 1.5-kg) bass is average on many waters. In general,

Largemouth bass often hide in weedy areas.

largemouths are more common the farther south one goes, while smallmouths are better suited to northern regions.

Largemouth bass love to lurk in weed beds, below lily pads, around tree stumps, and near rocky areas. That means these bass prefer fairly shallow, warm water. They eat other fish, frogs, crayfish, and bugs. Good artificial baits include spinnerbaits, chugger lures, and soft, plastic imitations of prey. Anglers use sturdy rods to help pull bass out of the weeds or cover before the fish tangles up the line.

Smallmouth bass are almost always found near rocky areas, where their favorite prey—crayfish— are often found. But minnows are common prey too. Smallmouth are famous for hitting hard and fighting even harder on the end of the line, often with spectacular leaps.

## CATFISH

Catfish are popular game almost everywhere they are found. Catfish bite hard and are good to eat, too. Though sometimes found in lakes, catfish are often considered more of a river fish. All catfish are predators to some extent. Some catfish also scavenge for food. Catfish have barbels, which look like whiskers and help the fish smell and sense prey or food. Catfish are nocturnal, which means they are most active and do much of their feeding at night. That's when many anglers go after catfish.

Channel catfish are sleek, streamlined, and silvery fish. A three- to five-pounder (1.5 to 2.3 kg) is a good channel catfish in most places. Channel catfish sometimes hunt for their food, while other times they are scavengers.

Flathead catfish are true predators. Flatheads are also called mud cats. A five- to 10-pound (2.3- to 4.5-kg) flathead is a good one, but flatheads can grow all the way up to 70 pounds (32 kg) or more. Anglers need heavy gear to haul them in.

Bullheads are another kind of catfish. Bullheads are smaller than flatheads and mudcats.

## PIKE

Fish within the pike family include northern pike and muskellunge. These top predator fish have lots of

Pike have sharp teeth for catching their prey.

sharp teeth. These predators mostly eat other fish, such as walleye, bass, and perch.

Northern pike are commonly found in and near weed beds. Here, pike lurk and wait in hiding, or patrol, for prey. Muskies are hard to catch, as there usually aren't many of them in any lake or river.

# BAIT

The key to fishing success is getting a fish to strike, hit, or bite the bait at the end of the line. Anglers have two main options for fishing bait: live bait and artificial lures. In some cases, anglers combine live bait and artificial lures.

## LIVE BAIT

Live bait is just that—natural items that fish like to eat. The live bait is put on a hook. One challenge to live-bait fishing success is keeping the bait healthy and lively so fish are attracted to it.

Nightcrawlers and worms have a good odor, and they wiggle enticingly to make fish interested. Nightcrawlers and worms need to be kept in a cool, moist place out of the sun.

Minnows are good bait because they are the natural prey of many fish.

A colorful artificial lure can be more effective if covered with fish-attractant scents.

Minnows make great bait because so many gamefish focus on smaller fish as prey. A minnow for crappie fishing might be an inch (2.5 cm) long, while a sucker "minnow" for pike fishing might be nine or 10 inches (23 to 25 cm) long. An aerated bait bucket and cool water keep minnows lively.

Leeches seem slithery and gross, but fish are attracted to their motion and scent. Leeches store well in a tub or bucket of cool, fresh water.

Grubs and larvae such as waxworms, spikes, euro larvae, and goldenrod grubs may be small, but fish love them. Grasshoppers and crickets work well as bait. Anglers can house them in a mesh box or tube.

## ARTIFICIAL LURES

Artificial lures imitate live prey. One advantage of artificial lures is they don't need to be kept alive. They don't smell like the real thing, but today's anglers often use fish-attractant scents to add odor to the color and action of artificial lures. Artificial lures can be cast and retrieved, jigged up and down, or trolled.

Spinners and spinnerbaits have blades that spin and flash. That attracts fish. Minnow baits and plugs imitate minnows. These baits are often designed to wiggle and shimmy like injured prey.

Chuggers and topwater baits ride on the surface of the water. They often have blades or scooped-face designs that gurgle water to attract fish.

Soft plastics are carefully designed and made to look like prey such as minnows or crayfish. When fish bite, the soft and realistic texture makes them hang on longer.

## TYING FLIES

Many fly anglers tie their own flies. Fly-tying is very precise and creative work. Materials used include thread, feathers, fur, tinsel, foam, beads, or yarn that is wrapped, tied, and sometimes glued onto a hook.

Flies imitate prey. Dry flies float on top of the water, and fish rise up to eat them. Nymphs and wet flies ride underneath the water's surface to imitate hatching bugs. Streamers mimic minnows. Poppers are cork- or balsa-bodied bug imitations that gurgle on top of the water.

Jigs feature a round or bullet-shaped head on a hook. Sometimes jigs are paired up with soft plastic tails, rubber skirts, or feather adornments. Many anglers tip their jig with live bait such as a minnow.

## FLY-FISHING

Fly-fishing is very popular. Trout are a traditional target of fly anglers. But any kind of fish can be caught using fly-fishing gear.

In fly-fishing, the reel holds a different kind of line that serves as the weight for casting. A leader of light line is tied to the end, and the fly is tied to that.

Many kinds of flies can serve as the lure. Dry flies and poppers ride atop the water, and fish slurp, swirl,

Many anglers like to tie their own flies.

or smack at them. Wet flies and nymphs imitate insects hatching in the water. Streamers mimic baitfish. *Terrestrials* is a big name for flies that imitate crickets, grasshoppers, ants, and other bugs that fall from land into the water.

# CHAPTER 6

# SAFETY TIPS

Fishing is fun. It's generally a pretty safe outdoor pastime too. But smart anglers know that fishing can be dangerous. The water itself is a potential hazard. Bad weather can brew up fast. And the process of casting sharp hooks on the end of a line can injure partners or an angler.

## LIFE JACKETS

Even anglers who can swim well should wear a life jacket at all times when fishing. Anyone can knock their head when falling out of a boat or hit their head on a rock when fishing from shore. A person who falls into a cold body of water can experience shock-like symptoms. A life jacket can keep someone afloat and

People of all ages need to wear a life jacket while on a boat.

hold their head above water even if the victim is hurt or unconscious.

## WEATHER WATCHING

Safe anglers are weather watchers. They keep their eyes on the skies for approaching storms, big winds, or other bad weather. If bad weather threatens, they quit fishing and get off the water fast. Lightning can strike anywhere. Big winds create waves that can overwhelm a boat, canoe, or kayak. Cold water is dangerous, even if an angler is wearing a life jacket. Smart anglers know that no fish is worth risking their lives for. "Live to fish another day" is an angler motto. And fish don't bite well in truly bad weather anyway.

Nobody wants to be caught on the water in stormy weather.

## CANOES AND KAYAKS

Canoes and kayaks are very effective fishing tools. They help anglers get onto small-water hotspots or back into out-of-the-way places where big fish swim. But several safety rules are important. A person should never stand up in a canoe. Canoes can be very unstable, and they tip over easily. Canoers and kayakers should stay fairly close to land so it's easy to kick to shore if they tip over.

Anglers need to be aware of their surroundings before casting a line.

## CASTING SAFETY

Many anglers like to hit the water with a friend. Fishing partners need to keep track of each other and always watch where their backcast goes so as not to hook their partners. That is a painful injury that is easily avoided with a little care and consideration.

Fishing is a sport that people of all ages can enjoy. It's often a tradition shared by friends and family members for generations. It's a great way to experience nature and maybe even feed the family. So grab a rod and reel, load up a tackle box, and try your luck!

# GLOSSARY

**aerated**
Provided with air or oxygen.

**angler**
A person who fishes with a hook as opposed to a net or a spear; also called a fisherman or fisherwoman.

**backwater**
An area of a river that is off to the side and out of the current or main flow.

**cast**
The process an angler uses to toss out the bait or lure into the water.

**current**
The continuous flow or movement of water in a certain direction.

**dike**
A long wall or embankment built to prevent flooding.

**eddies**
Circular movements of water, counter to a main current, that create a small whirlpool.

**forage**
Another word for the food fish eat, also called prey.

**habitat**
The physical attributes of an area where fish, plants, or animals live.

**landing**
The process of hooking a fish, fighting it, and pulling it in.

**predator**
A fish that eats live prey.

**scavenger**
An animal or fish that eats dead things rather than hunting and killing its own food.

**trolling**
The fishing process of rowing or motoring slowly along in a boat and pulling a lure or bait behind.

**waterfowl**
Ducks, geese, or other large aquatic birds, especially when regarded as game.

# MORE INFORMATION

## BOOKS

Lukidis, Lydia. *Fisheries*. New York: Smartbook Media, 2018.

Maas, Dave. *Kids' Guide to Fishing: The Young Angler's Guide to Catching More and Bigger Fish*. Mission Viejo, CA: Walter Foster Jr., 2018.

McDowell, Pamela. *Fishing Village*. New York: AV2 by Weigl, 2016.

## ONLINE RESOURCES

To learn more about fishing, please visit **abdobooklinks.com** or scan this QR code. These links are routinely monitored and updated to provide the most current information available.

# INDEX


backwaters, 13
bait, 20–21, 31, 34–39
bass, 12–13, 17, 23,
     29–31, 33
benefits, 6–7
bluegills, 28
boats, 22, 40–43

catfish, 13, 23, 31–32
crappies, 28, 36
creeks, 12, 17, 26

fly-fishing, 38–39

hooks, 23

knots, 18–19

lakes, 8–11, 22, 26–28
line, 18–19, 21, 38
lures, 20, 31, 37–38

muskies, 17, 21, 32–33

panfish, 12, 17, 23,
     27–28
perch, 33
pike, 21, 23, 32–33, 36
ponds, 11, 17, 26–27

reels, 14–17, 38
rivers, 12–13, 22, 26,
     28, 31
rods, 14, 17, 31

safety, 40–45
salmon, 17
saugers, 28
sunfish, 28

tackle, 19–22
trout, 12, 17, 23,
     26–27, 38

walleye, 12, 17, 23,
     28, 33


# ABOUT THE AUTHOR

Tom Carpenter is a father, a sportsman, and an outdoor writer. He has introduced many children, including his three sons, to the thrills and rewards of fishing. A native of Wisconsin who always has part of his heart in South Dakota, he planted roots in the middle. He lives with his hunting dog, Lark, near the shores of Bass Lake, Minnesota.